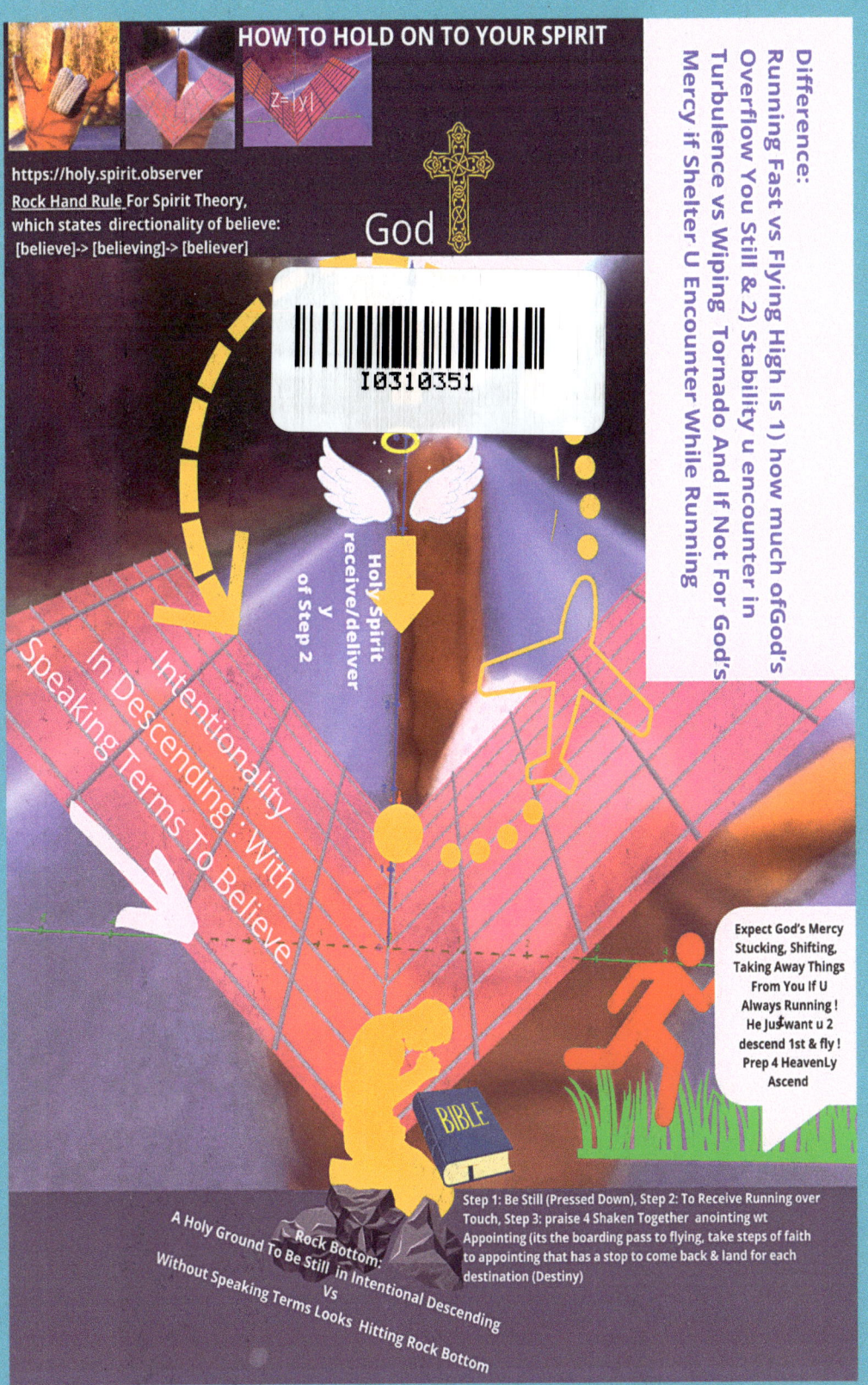

Book References & Editions
https://holy.spirit.computer

Copyright © 2022
ISBN : 978-0-578-26579-7
All rights reserved. No portion of this book may be reproduced in any form without permission from the publisher, except as permitted by U.S. copyright law. For permissions contact: creations@spirit.institute

SPIRIT PRESS℠
https://holy.spirit.help
2022

Spirit Science | Observation & Theory
AT, SPIRIT ECONOMY INC.
A **GOD**-1st IP Creation, Commerce & Std. Co.

How To Read
SPIRIT COMPUTER 1.0

OPERATING NEURAL WIRE USING LEAP OF FAITH ALGORITHM (METHOD)

Delight Steps
Gear 1 - Now
Gear 2 - One Step Back : Bible, Resting, Set Still Sight
Gear 3 - Two Steps Forward: Angles Convert Obstacle To A Cart & Blow You On It Thrusting You To A New Level Of Mercy To Get Wired While At Still & Joy
Gear Stuck - Staying Long, Sunk, Cornered -With Out Bible

Desire Steps
Gear 1 - Now : Carrying Baggage & Scar of Past, Hard To Smile (p.40)
Gear 2 - One Step Back With Out Bible : Unkind Pressing Pressure
Gear 3 - Two Steps Forward: Manually Elongating By All Body Will, Weary, With Worry, Never Resting
Gear Stuck - Good Thing You're On Fight Mode Never To Get Stuck, But Eventually To Come Here When You Get Weary By #3 At Early Age Being Always On The Run

**BE ENCOURAGED
SPIRIT
IS A FACULTY FOR SELF HEAL**

Disclaimer

Spirit Computer is a computational model, an explainer to establish speaking terms (via John 16:13-14) by the language of your Spirit: SIGHT. Contexts in this book are taken from a Biblical inspiration to allow you to trust the built-in faculty of your spirit in order to be living your life through it (Romans 8) computationally and effectively for your wellness, for your excellence, for your relationship with one another, and most importantly for your daily resurrection power activation!

 Value of life : #1 of page 56 & 58 inserted into page #16 life pursuits

SPIRIT COMPUTER 1.0 OUTCOME : GET TOUCH BY SIGHT
- SENDING BY YOU: SET SIGHT TO GOD (DAILY SUBSCRIPTION RENEWAL OF YOUR SPIRIT)
- FULFILLMENT BY: BY CHRIST THRU HOLY SPIRIT (A JOY & PEACE IN HIS ABILITY, SO SHOW HUMILITY)
- RECEIVING BY YOU: TOUCH (CALLING, HEALING, CLARITY...) (SURRENDER YOU WILL WITH GOD'S)

 SHOW RESPECT TO YOUR SPIRIT: NO GASLIGHTING! LISTEN WITH A GOD 1st GUT, IN YOUR SPIRIT, WITH HUMILITY FOR TRUTH (JOHN 16:13)! LISTENING TO GOD'S WORD FOR TRUTH AGAINST LIES IS THE ACTIVATION OF THE TRUE ADVOCACY POWER OF YOUR SPIRIT! INVITING GOD TO BE 1st IN YOUR LIFE ALLOWS HOLY SPIRIT, YOUR ADVOCATE, TO DWELL IN YOU! FOR A RELATIONSHIP! YOUR ADVOCATE ONLY SAYS TO YOU WHAT IS RECEIVED FROM JESUS CHRIST (JOHN 16:14)! WITH THIS RELATIONSHIP, YOU CAN OPEN AN INVITATION FOR GOD'S LIMITLESS TOUCH TO TRANSFORM YOUR LIMITATIONS: MOVE YOUR GIANTS, FLIP DYSFUNCTIONS TO GREATNESS WITH GIFTS', CURE AILMENTS & MOST IMPORTANTLY RECEIVE THE CALL FROM OUR LIVING CHRIST LIKE LAZARUS, TO "COME OUT!" FROM ANY OF OUR SITUATIONS & DEATH (JOHN 11:43)!

አሜን!

Preface

Lord, Thank You for Imparting On Me The Courage To Write, To Develop This Content! Your Holy Spirit Was Holding My Hands To Create The Contents In This Book! I Didn't Even Plan On This To Be A Book, Until Just About A Few Minutes Ago, But You Planned It For Me, Turning A Flyer To A Book! May The Contents In This Book, Showcase The Methods For Us To Dwell On Your Words Found In Our Bible, For Us To Be The Generation & Bloodlines In Your Best Plan Striving For Salvation, Welcoming Your Return, To Be Filling Your Heaven By Us Bearing Fruit Here On Earth, 100x As In Mark 4:13, In Our Believing As In John 6:29, Through Jesus Christ The Author Of Our Faith, Who Imparted On Us The Holy Spirit (John 14:15, Luke 3:16)!

Grateful To My Ever Supportive Parents, Momishaye, Aster, Our Family's Rock, Who Showed Me The Cornerstone (Matthew 21:42), And Babishaye, Our Enduring Kind Father, Eng. Zelalem, Who Instilled The Joy In Being Fruitful (Matthew 25:20), & To My Siblings : Abishaye, My Inspiration To Holding On Firm To Shielding Your Grace (Eph. 6:13) & Joye, My Inspiration of Obedience (Eph. 6:1), To My Overlooking Angels My Family & Friends Far & Near, Especially My Church Small Groups, & My Virtual Paster, Ps Joel Osteen & Community In My Upbringing of Believing To Finding God Through The Regimen of Ethiopian Tewahdo Church! I Fall Short To Be Writing This Book In The Eye Of Not Only My Heavenly Father But To All Mentioned Circle Of God Set Supportive System!

Gift

God has your gift buried,
Just so you seek his touch
to see His miracle of,
raising from the buried
to planted, to blooming!
The Fruit of Heavenly Living!

Afterwards, It won't be about the desires of your heart in the net breaking fish, but about the miracle set in motion, by His touch! Like Peter, walking out of his desires to catching a fish & walking into to the delight of the Lord, in which you don't want your dear ones not missing this heavenly touch!

Your Gift Transformation Is An Early Confirmation Of A Resurrection !

about

I lost everything or opportunity for everything :(but it was for a relationship...to be connected...to establish speaking terms...to Who owns everything! Like Peter, I keep launching the net, but then I keep recoiling lacking coherence, realizing it wasn't my calling or the best bet to be at rest! Like Peter, the finding of the Owner of the ocean, the Creator of The Universe, became my permanent glory, now telling you here in this book that your best is found in the touch after establishing speaking terms with your spirit causing you to rest! What comes next is endurance work, but on the foundation of that same Touch that sets miracles in motion for coherence while still remaining at rest! That truth! That truth is a dimensional change! A foundational test! The truth or lie test to know what's in it for the best of your destiny has not changed since Adam & Eve. You pass it with your spirit. Computing. Truth to be told is you have to be formed to know the truth from the lie, before the shakeup (Luke 21:34-36), discerning! Visit https://2136.life for this lifestyle devotional! Found in your spirit, you want not but need to be connected to your faculty of conviction, your spirit, to receive the forgiving love of God's touch, for an unequivocal hunch: equipping you with the knowledge of how destined you are to fly high, at the right foundation (John 14:6)!

kirubel zelalem seifu

GOD | ecosystem
1st | inventor

TWIST

IF YOU DON'T HAVE A FIST
HOLDING GOD'S STORED BEST
U BECOME FIT FOR A TWIST
WANT SWITCHING IN FOR NEED
NEED SWITCHING IN FOR WANT
A LIVING UNFIT IN THRIVING
A LIVING FIT FOR SURVIVING
GOD BRINGS IN SITTING
NOT LEAVING UR BELIEVING
GOD MADE THRIVING A NEED
NEED IS A GOD SIZED DEED
BLESSINGS BY GOD ONLY INDEED
AS GOD IS A LIVING GOD IN NEED

Contents

SPIRIT POWER ON 1
SPIRIT OPERATING SYSTEM   11
SPIRIT SEED BOARD   15
SPIRIT PROCESSING UNIT   17
READ ME   69
FAQ   71
SPIRIT POWER OFF   73

1

**Believe
(Circularly From
Believe, Believing, Believer)
Is The Work of Life
John 6:29**

Intro

Believer — 3. Step 1 is Here, Praise when Shaken Together (Both: Shaken Wt Company of Love Wt Joy & Shaken together of 4 received Touch in #2 anointing, for soaring delight steps of Appointing) *Luke 17:11-19*

2. Live Life: Thru Faithfiled Depth Of Surrender (Its A Grounding Strength Wt God 4 Receiving Touch)

Believing —
Depth of Surrender In Love **(2.1) First** **& Excellence 4 Running over Abundance** **(2.2) Mark 4:20**

Believe — 1. Be Still: 1.1) Pressed Down /Stay 1.2) Get in Agreement Wt God Laying Foundation 4 Holy Spirit Dwelling 1.3) So as To Receive Anointing of Honoring God In Wispering 4 Living in Delight

```
Psalm 46:10, Be Still
Knowing
isaiah 6 commissioning
By
Jesus John 16:14
For
God John 14:21
```

↑ Fruitful
↑ SPIRIT COMPUTER
↑ Faithful

Breath in! God's Word
Breath Out! God's Word

By the word of the Lord were the heavens made; and all the host of them by the breath of his mouth - Psalm 33:6

Be Still, And Know That I am God

BE STILL
PSALM 46:10

2

SPIRIT COMPUTER

WHAT For **Bible**
YOU Neural
NEED Wiring

Pen Paper

or Notepad

Spirit Computer Is An **Instant, Day, & Lifespan** Coherence Computing Model To Be Still, By Putting God 1st, Pouring **In** & **Out** God's Words !

The reality is God has already breath out life on us (Ps 33:6), meaning we can call in His promises now, throughout our day & lifetime & be still! Being still has its way more than through the breathing in deep & out! In this self-help therapy for establishing speaking terms, a successful test result hypothesizes that you can be still, through a sighting of a God's 1st state of mind, whereby the **neural wiring to a God-1st brain state is unique than otherwise!**

https://holy.spirit.computer

⃝ spirit.computer

PARABLE
ANTENNA

Breath in! God's Word
Breath Out! God's Word

By the word of the Lord were the heavens made; and all the host of them by the breath of his mouth - Psalm 33:6

Be Still, And Know That I am God

BE STILL
PSALM 46:10

SPIRIT ?

Seeing Spirit As Your
Directional Antenna
Meant To Recieve
Holy Signal

There are different kinds of antennas. but directional antenna is the one you have to point to a sender, like how you hold on to a flash light towards a target to get a clear vision!

PARABLE
NO DATA

Breath in! God's Word
Breath Out! God's Word

By the word of the Lord were the heavens made; and all the host of them by the breath of his mouth - Psalm 33:6

Be Still, And Know That I am God

BE STILL
PSALM 46:10

SPIRIT ?

Unlock

Your phone receives data (content), upon **REQUEST** but not when isolated or locked or airplane mode! Unlock for today's grace thru the WORD of God in praise, prayers like your phone to receive today's grace

ENOUGH

Breath in! God's Word
Breath Out! God's Word

By the word of the Lord were the heavens made; and all the host of them by the breath of his mouth - Psalm 33:6

Be Still, And Know That I am God

BE STILL
PSALM 46:10

THE OBEDIENCE
THE BELIEVING
THE SHINING

HOLD UPRIGHT LIKE FLASHLIGHT

The beauty in life is in the intentional handling of relationships : which is a management of speaking terms by your spirit between the bird eye view of life (God's view) & worm eye view of life (yourself in your pursuits)!

In order to be getting signal, obedience is in holding onto your antenna towards the source.

The believing is the knowing of a wifi tech ability that it can beam signal without you seeing the signal in the thin air by your eyes but still connecting onto your devices.

See flashlight as an antenna, the beaming light as a wifi! Holding it upright in a dark room, it has a wide enough beaming, to break off from a dark sight, instead of holding it to the side or down, which gives a linear, worm eye view at a targeted sight.

The shining is in believing there is enough grace, it streaming through your turning on of your wifi -- by surrendering; letting go, letting God! Like how your wifi beams out from source far high!

UNBELIEVING

By the word of the Lord were the heavens made; and all the host of them by the breath of his mouth - Psalm 33:6

Be Still, And Know That I am God

BE STILL
PSALM 46:10

UNBELIEVING

CIRCUMSTANCE

Unbelieving is like extending a cord to external router manually to someone's else antenna endpoint! You have to manually do it all the time! You may have to pay for content you are fed. You may get free wifi to connect using your wifi, but it's not always easy to get that! That's usually when you grow up under your parents who pay for their wifi ! Afterwards, once a while you may get free runs but that connection isn't strong signal to install big programs which help you in accomplishing big milestones, such as streaming entire content for your exams to get your degree! Unbelieving could be doubting its cloudy and that your antenna wont work! It's discounting God when he created cloud, as if God failed to understand your design standard !

SPIRIT OPERATING SYSTEM

Be Still, And Know That I am God
BE STILL
PSALM 46:10

By the word of the Lord were the heavens made; and all the host of them by the breath of his mouth - Psalm 33:6

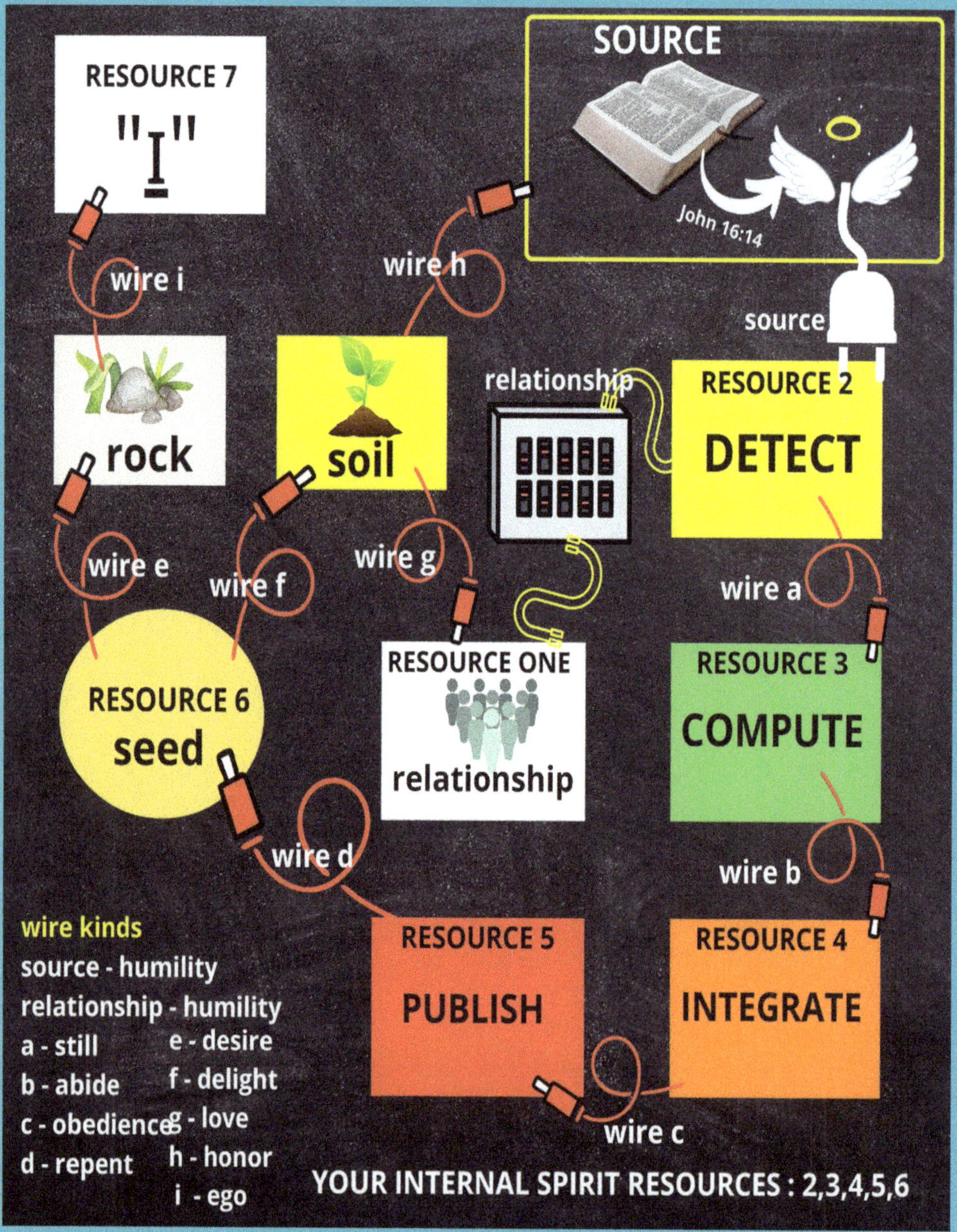

RESOURCE 7
"I"

SOURCE
John 16:14

wire i
wire h
source

rock
soil
relationship

RESOURCE 2
DETECT

wire e
wire f
wire g
wire a

RESOURCE 6
seed

RESOURCE ONE
relationship

RESOURCE 3
COMPUTE

wire d
wire b

wire kinds
source - humility
relationship - humility
a - still   e - desire
b - abide   f - delight
c - obedience g - love
d - repent   h - honor
             i - ego

RESOURCE 5
PUBLISH

RESOURCE 4
INTEGRATE

wire c

YOUR INTERNAL SPIRIT RESOURCES : 2,3,4,5,6

DETECT FORGIVING LOVE

More on forgiving love course
https:// forgiving.love

Heavenly sight (Luke 8:15): Hear, Retain, Persevere
Earthly Sight (Luke 8:18): Soil, Pathway, Rock, Thorn
Earthly & Heavenly Intersection: Be Still, on Soil

GOAL OF FL Course is to be in seeking of forgiving God's love, so you can be sowing your life efforts on soil by showing forgiving love Chirst showed you!

PS: To Be Still, encounter each level, from hearing to persevering (which is a believe in depth/faith)

BE STILL
Psalm 46:10

Heavenly Sight
- Hear
- Retain
- Persevere

Earthly Sight: Soil, Pathway, Rock, Thorn

Path. Rock. Thorn. Soil

Path: Heard About God But Ain't Believing 2 Be Saved
→ Holding God 1st Loosely
- Believing in God On limited conditions!
- Inconsistent Praise & Prayer !
- Taking out God On Fruition of Gift Relying on Strength
- socials Bad company Contamination

Rock: No moisture To retain seed
→ (Relapse Zone)

Thorn: Not maturing Choked By Worry, Riches, Pleasures
→ Hardened Heart Towards Another & God

Soil: Nobel & Good Hearted
→ Loving One Another
(Not by righteousness But By Seeking God's Forgiving Love)
→ God 1st

DETECT
RESOURCE 2

Spirit Computer 1.0 is based on Spirit Theory. Spirit Theory explains wellness in forgiving love. (p.49)

Spirit Theory is Forgiving Love.

Parallels

Believe > Believing > Believer

God > Love > Gift

ONE STEP BACK > Now > The Next

Overcoming
BY DETECTING

STEP 1 : Environment
- Social Contamination/ Distractions

STEP 2 : Thru God 1st
2.1 : 1st, Heart (In Spirit) : Unbelieving God Speaks Here In Holy Spirit
2.2 : 2nd, Will : Letting 2.1 Overcoming Desires
2.3 : 3rd, Mind : Now Your Still Unlocking Discipline 4 Gift

STEP 3 : Love
- Loving One Another

As In Ordered Steps Matthew 22:37

By the word of the Lord were the heavens made; and all the host of them by the breath of his mouth - Psalm 33:6

Be Still, And Know That I am God
BE STILL
PSALM 46:10

WIRED

- Getting wired down under your spirit is for letting your believing flow THE GRACE you need
- See the wire kinds (a-i) as seeing the amount of energy & strength different wires carry for varying needs
- See the wire types, inside wire (S-Type, B-Type, T-Type) as carriers of God's Word, in your content, to execute daily tasks

> S-Type For Prayer : Source wire, to call in God's Word, to fight your oppostion
> B-Type For Believe : Beleive to march on towards God's Throne & recieve your mercy, by believing that you are built to overcome your circumstances (John 16:33), in the progressive design of your faith! Be watchful against traps of :
a) ego -- this courage isn't to be boastful
b) decay -- against desires, personally & agianst social decay contaminations (as simple as through fashion, to excess consumptions) collectively
> T-Type For Praise : Give thanks for His breath of life on you, for fighting your battels for you, for holding His promise of mercy secure

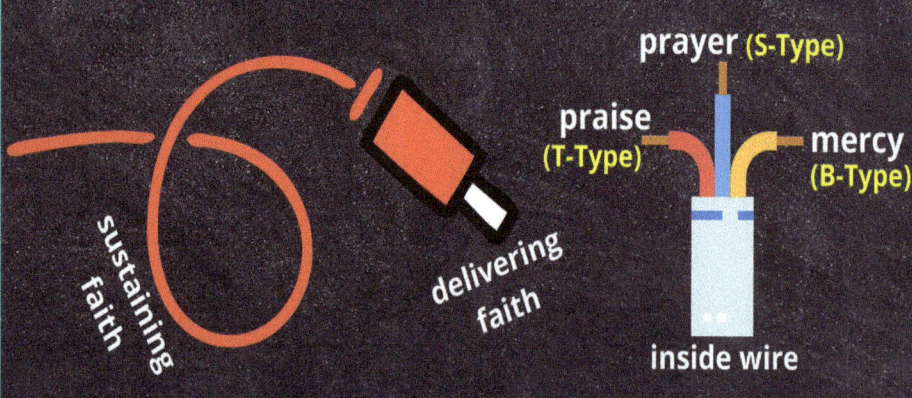

SPIRIT
SEED
BOARD

Breath in! God's Word
Breath Out! God's Word

By the word of the Lord were the heavens made; and all the host of them by the breath of his mouth - Psalm 33:6

Be Still, And Know That I am God
BE STILL
PSALM 46:10

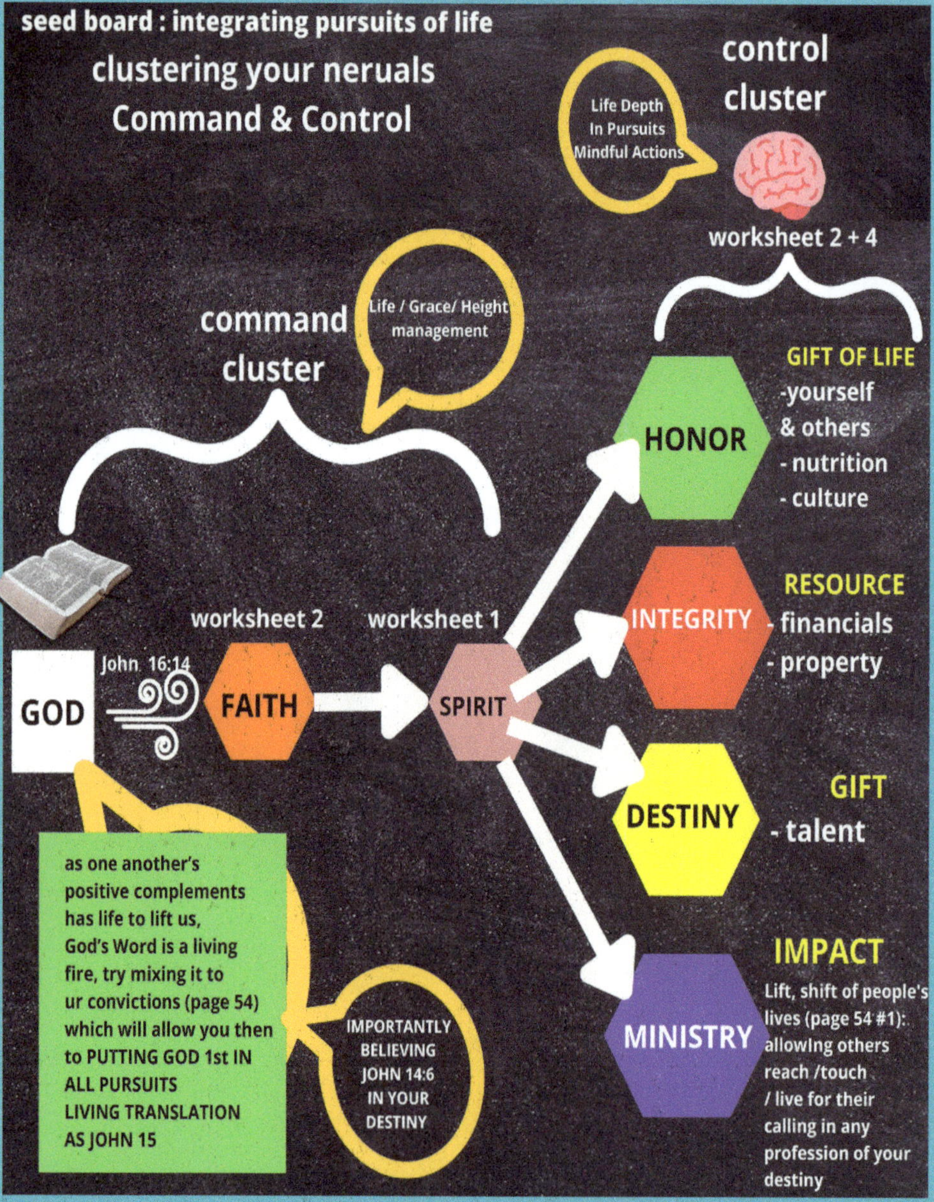

seed board : integrating pursuits of life
clustering your neruals
Command & Control

control cluster — Life Depth In Pursuits Mindful Actions

command cluster — Life / Grace/ Height management

worksheet 2 + 4

GIFT OF LIFE
- yourself & others
- nutrition
- culture
→ HONOR

RESOURCE
- financials
- property
→ INTEGRITY

GIFT
- talent
→ DESTINY

IMPACT
Lift, shift of people's lives (page 54 #1): allowing others reach /touch / live for their calling in any profession of your destiny
→ MINISTRY

GOD — John 16:14 → FAITH (worksheet 2) → SPIRIT (worksheet 1)

as one another's positive complements has life to lift us, God's Word is a living fire, try mixing it to ur convictions (page 54) which will allow you then to PUTTING GOD 1st IN ALL PURSUITS LIVING TRANSLATION AS JOHN 15

IMPORTANTLY BELIEVING JOHN 14:6 IN YOUR DESTINY

SPIRIT PROCESSING UNIT

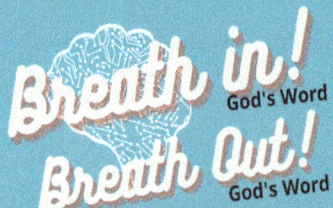

By the word of the Lord were the heavens made; and all the host of them by the breath of his mouth - Psalm 33:6

Be Still, And Know That I am God
BE STILL
PSALM 46:10

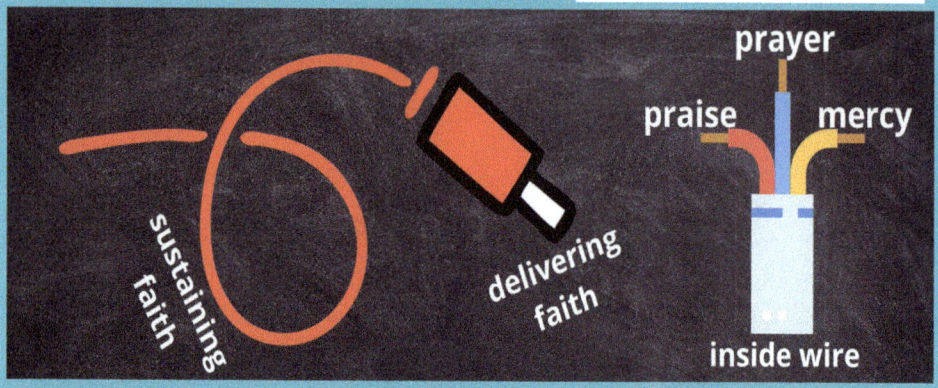

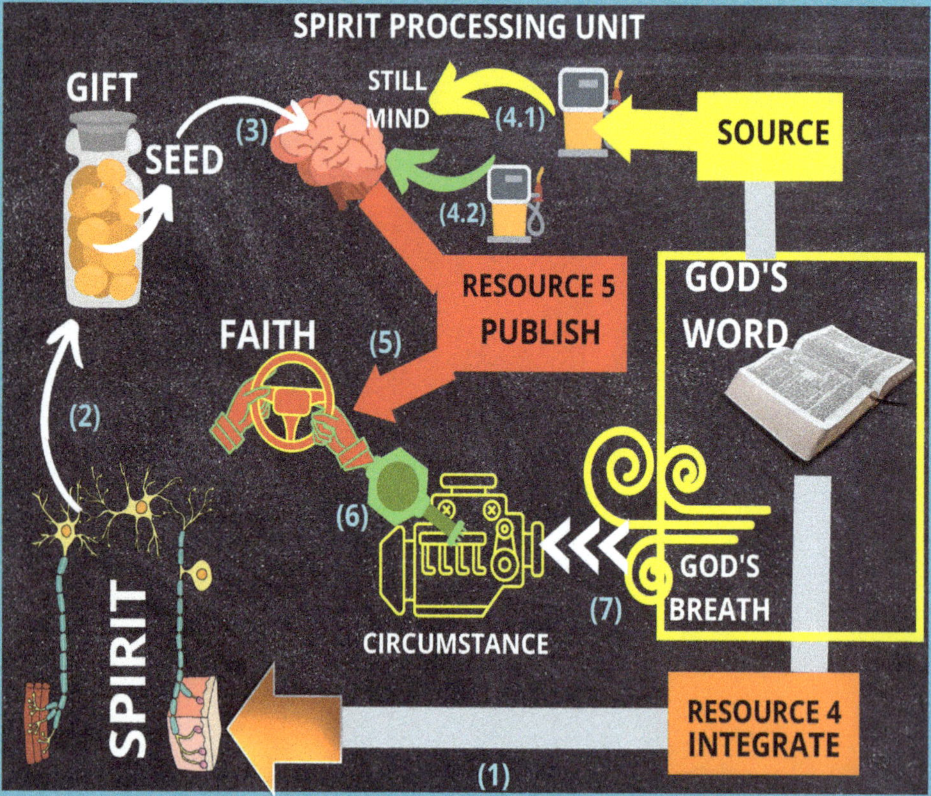

SPIRIT PROCESSING UNIT
GRACE FLOW

By the word of the Lord were the heavens made; and all the host of them by the breath of his mouth - Psalm 33:6

Be Still, And Know That I am God

BE STILL
PSALM 46:10

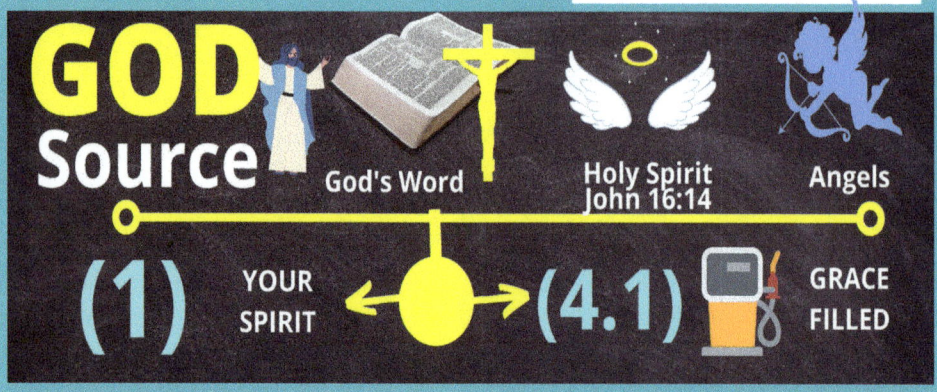

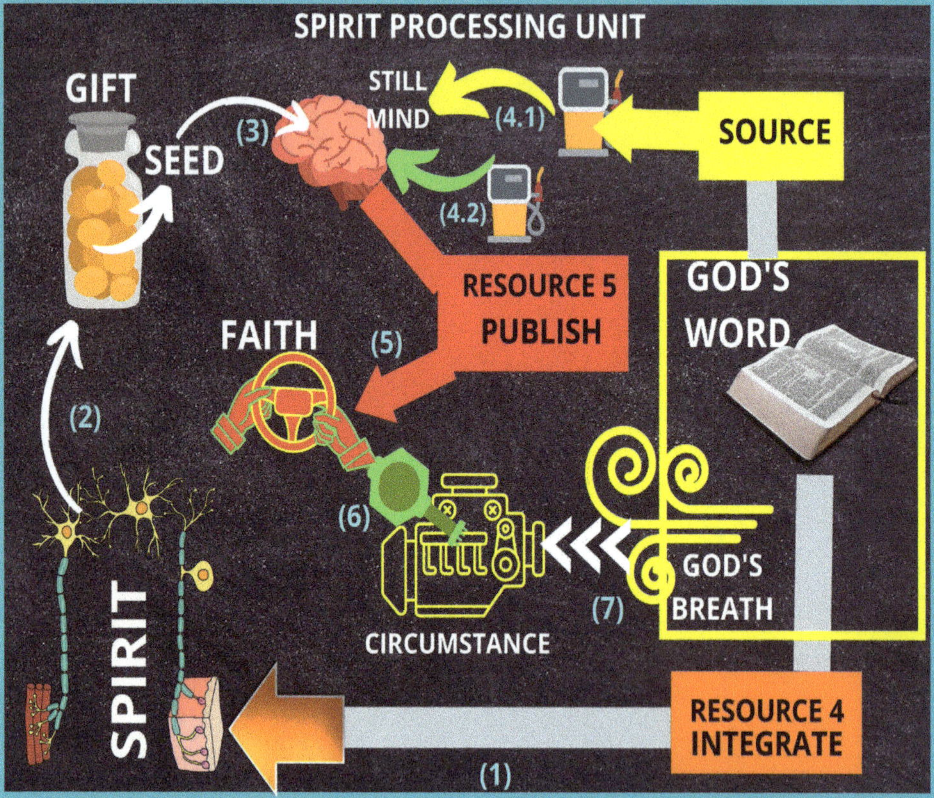

SPIRIT PROCESSING UNIT
DATA FLOW WORLDLY LOGIC

(4.2)

SG-40

LET INVENTIONS (PRODUCTS, SERVICES, CONTENTS, CREATIONS) FROM FASHION, MUSIC, TECH, CAR, HOUSE DESIGNS, LET COME OUT BUT HAVE THEM PASS THE SG-40 COMPLIANCE: HAVING THE GRACE SET TO CAUSE YOU TO BE STILL & SURRENDER NOT CORNERED WITH DESIRE, LEADING TO EGO & SELF-FULFILLMENT.
SG-40 : CONSUMPTION OF DATA NEEDING 40 DAYS OF FASTING & PRAYER AFTERWARDS. SERIOUS TO DERAIL YOU!

Breath in! God's Word
Breath Out! God's Word

By the word of the Lord were the heavens made; and all the host of them by the breath of his mouth - Psalm 33:6

Be Still, And Know That I am God
BE STILL
PSALM 46:10

DATA Source

(4.2)

Referring to A Repository of A Worldly Logic Built By Consensus Of Mainstream Culture! But It's Mixed, Unchecked Data, Requiring Maturity To Discern If Its Fruit Is Built On The Grace of **GOD SOURCE**

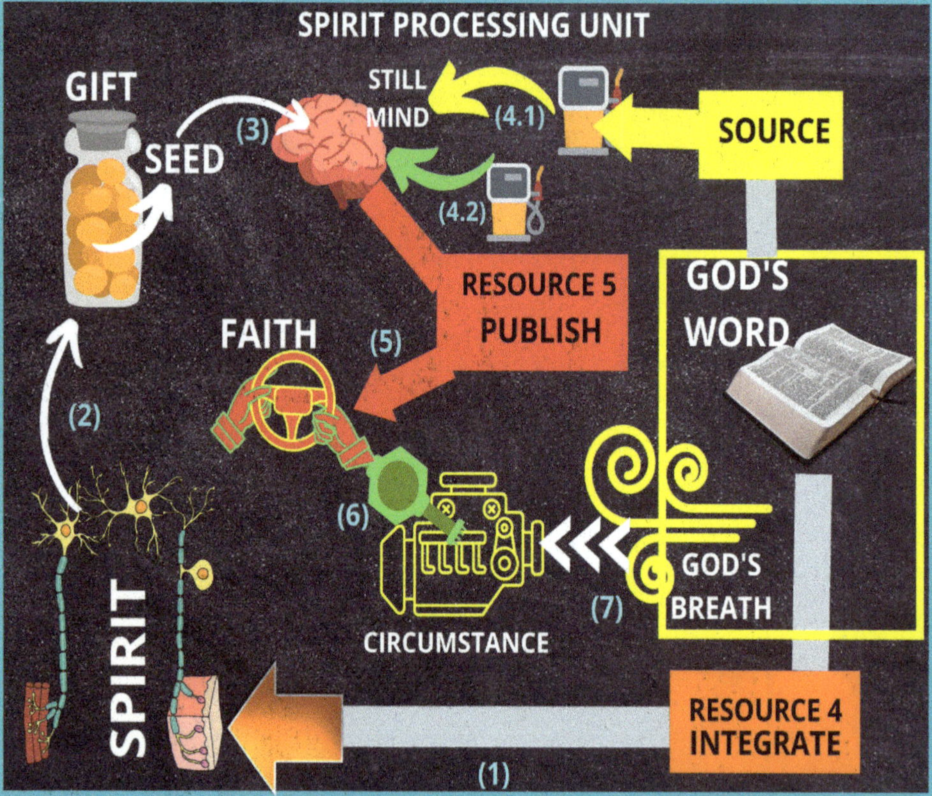

SPIRIT PROCESSING UNIT

- GIFT / SEED
- STILL MIND (4.1)
- (4.2)
- (3)
- SOURCE
- RESOURCE 5 PUBLISH (5)
- FAITH
- GOD'S WORD
- (6)
- CIRCUMSTANCE
- GOD'S BREATH (7)
- (2)
- SPIRIT
- (1)
- RESOURCE 4 INTEGRATE

SETTING SIGHT IN YOUR SPIRIT

COMMAND & Control (THE EFFECT)

LIFE ACTIONS ON BELIEVED SIGHT IN YOUR MIND

SPIRIT PROCESSING UNIT
FAITH

Even a **wink command** is enough to fly a drone! Your spirit likewise can receive a command that passes your understanding: a **divine touch** causing you to shine, fly high!

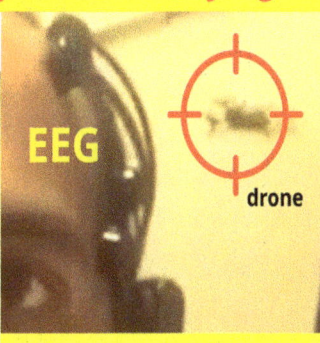

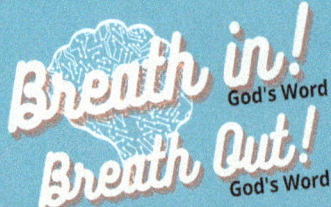

By the word of the Lord were the heavens made; and all the host of them by the breath of his mouth - Psalm 33:6

Be Still, And Know That I am God
BE STILL
PSALM 46:10

#3 Recall when said listen to your gut! Or your hunch!

Command & Control
Both Are Not in Your Brain

Command is in your Spirit (3)

Control is in your Brain

(2)

(4)

By Holy Spirit 4 Command

By Work 4 Control

(1)

(5)

God's Living Word In Holy Sprit

Grace in guiding logic 4 work

(6)

Fruitfulness By Grace
🍓🍒 🥑 🍎 🍌 🥥 🍇 🍉
🥝 🍍 🍋 🥥 🍏 🥭 🥝 🥥

FATIH: The Command & Control

SPIRIT PROCESSING UNIT
GROWING IN FAITH

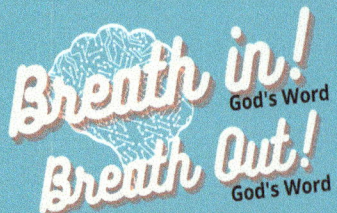

By the word of the Lord were the heavens made; and all the host of them by the breath of his mouth - Psalm 33:6

Be Still, And Know That I am God

BE STILL
PSALM 46:10

GROWING

Your Word

Your Creations

Call It In, It's A Believing In The Now **Subscription** of Your Sprit in **God's Power**

4 Putting God 1st
4 Healing
4 Surrender
4 Joy

Write /Declare it like this
I call in putting God 1st in all my ways
I call in for healing in my body
I call in for joy ....keep going!

Blend it in, In The Design!

The Product, Design, Result

God's Word

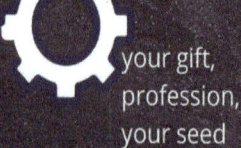

your gift, profession, your seed

SPIRIT PROCESSING UNIT
GROWING IN LIVING TRANSLATION

By the word of the Lord were the heavens made; and all the host of them by the breath of his mouth - Psalm 33:6

Be Still, And Know That I am God
BE STILL
PSALM 46:10

GROWING

to grow is ...
1) to praise you are now standing on the droplets of God's breath of life
2) to seek His blowing wind of mercy to continue to be thrust forward on His breath of life
3) to learn obedience which allows us to surf on His breath of life

---

praising we are given His breath of life to seek His delivering mercy where we show our growing obedience in order to be restored heavenly!

SPIRIT PROCESSING UNIT
GROWING AFTER TRANSFORMATION

By the word of the Lord were the heavens made; and all the host of them by the breath of his mouth - Psalm 33:6

Be Still, And Know That I am God
BE STILL
PSALM 46:10

GROWING

Discipline

while carrying along others to touch your new high, the strength you are gaining by converting others set back & pain with the word of God, has to be sowed by setting apart a time for fruitfulness & the Word of God you are shielded needs to be renewed with a community of believers in God's word who are in mentorship and in agreement with you, by this discipline your destiny is protected from contamination, and fruitfulness sowed ! Carrying along isn't giving unfettered access, but showing how to fish, giving the discipline for self-help!

SPIRIT PROCESSING UNIT
GROWING FOR HARVESTATION

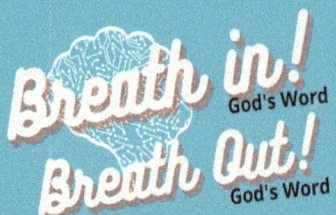

By the word of the Lord were the heavens made; and all the host of them by the breath of his mouth - Psalm 33:6

Be Still, And Know That I am God
BE STILL
PSALM 46:10

GROWING

Super

Are you doing the natural, from
n to **a** to **t** to **u** to **r** to **a** to **l**
so God breath out His Super for a supernatural increase! His Divine, His Super breath as a beginning to your life, needs to also continue in your natural cultivated gift, your gift is God sized, beyond your strength for all creations are supernatural
Call in for Super!

SPIRIT PROCESSING UNIT
GROWING FOR FRUITFULNESS

By the word of the Lord were the heavens made; and all the host of them by the breath of his mouth - Psalm 33:6

Be Still, And Know That I am God
BE STILL
PSALM 46:10

GROWING

Fruitful Will

Your **WILL**—is disgned to dig its heel in with strength & be **FRUITFUL** when it's used in **BELIEVING** for receiving in thankful steps or still stops from putting God's breath **FIRST** & from giving God the **FINAL** Say!

34

SPIRIT
PROCESSING
UNIT
GROWING
STEADY OR NOT

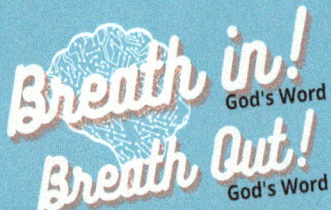

Breath in! God's Word
Breath Out! God's Word

By the word of the Lord were the heavens made; and all the host of them by the breath of his mouth - Psalm 33:6

36

Be Still, And Know That I am God

BE STILL

PSALM 46:10

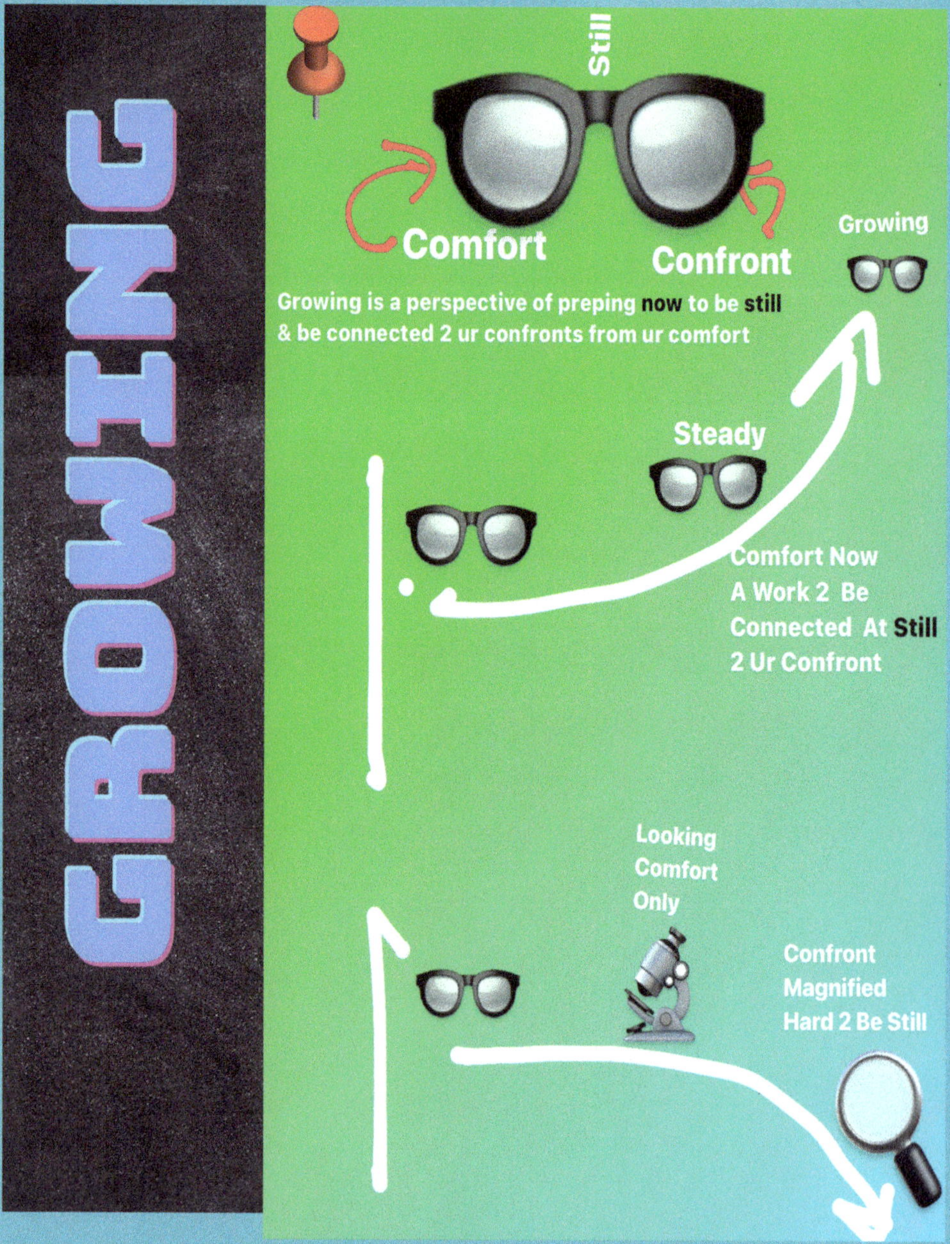

GROWING

Still

Comfort — Confront — Growing

Growing is a perspective of preping **now** to be **still** & be connected 2 ur confronts from ur comfort

Steady

Comfort Now A Work 2 Be Connected At **Still** 2 Ur Confront

Looking Comfort Only

Confront Magnified Hard 2 Be Still

COMPUTE
FRUIT OF SPIRIT

God's Word

God's Word

By the word of the Lord were the heavens made; and all the host of them by the breath of his mouth - Psalm 33:6

Be Still, And Know That I am God

BE STILL
PSALM 46:10

CONNECTING THE DOTS TO
COMPUTE
FRUIT SPIRIT

setting sight on desire manifests a feel in the burning of flesh!, causing wear! Setting sight on delight manifests a nourishing feel of a river of life : THIS RIVER OF LIFE IS THE JOY REFERRED HERE, SETTING SIGHT IN GOD 1ST. A HEALING JOY

JOY  THEY ARE CONNECTED 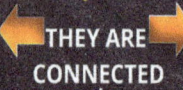 CHARM	
IS AN INSIDE OF YOUR SPIRIT	IS A REFLECTION ON THE OUTSIDE

Let's Compute The Spirit of **SMILE**

GOOF	REAL	FAKE
You didn't mind to understand	Mindfully Reasoned why	Because it is not conneced to fruit of joy

COMPUTE
MEDICAL THERAPY

Breath in! God's Word
Breath Out! God's Word

By the word of the Lord were the heavens made; and all the host of them by the breath of his mouth - Psalm 33:6

Be Still, And Know That I am God
BE STILL
PSALM 46:10

DIVING DEEP INTO THE BIOLOGY TO
COMPUTE
THE MEDICAL THERAPY BY YOUR SPIRIT

JOY ← THEY ARE CONNECTED → **CHARM**

IS AN INSIDE OF YOUR SPIRIT | IS A REFLECTION ON THE OUTSIDE

MEDICAL THERAPY BY YOUR SPIRIT

FRAWN | **SMILE**
62 MUSCLES | **26 MUSCLES**

DUNCHENNE SMILE (SCIENTIFIC NAME)
REAL SMILE; REAL = JOY+CHARM

The two muscles that create a smile are the **zygomatic major** and the **orbicularis oculi**

COMPUTE
COHERENCE CHART

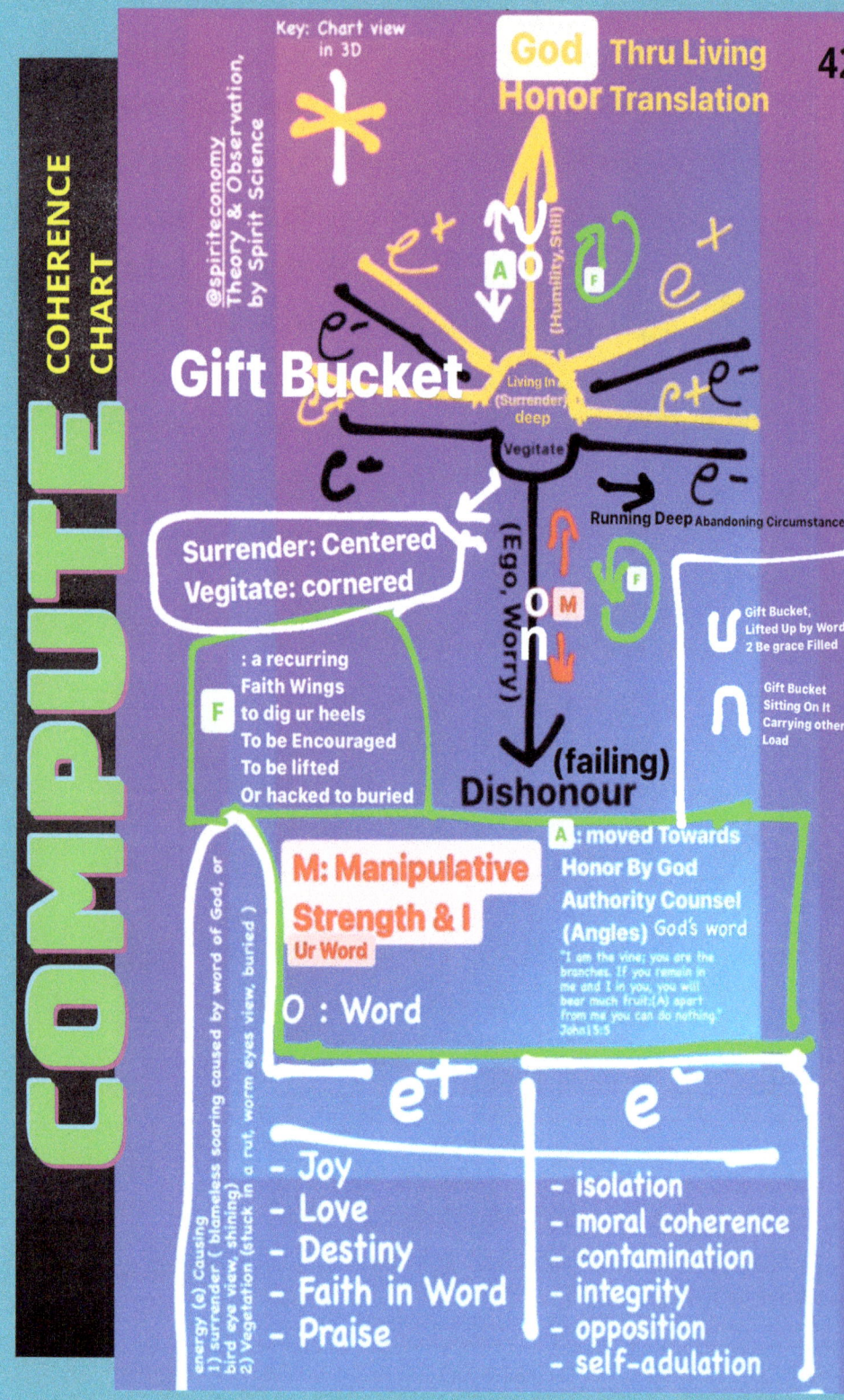

COMPUTE
HACKED MIND

44

Breath in! God's Word
Breath Out! God's Word

By the word of the Lord were the heavens made; and all the host of them by the breath of his mouth - Psalm 33:6

Be Still, And Know That I am God
BE STILL
PSALM 46:10

NO GOD 1st STATE OF MIND, NEEDS A
COMPUTE
TO RESOLVE A HACKED MIND FROM ITS SIGHT (IN SPIRIT)

URSELF, THINGS, SOMEONE

LIE NEEDS GASLIGHTING TO SHIFT SIGHT

Spirit

Gaslighting is After Your Spirit So Ur Mind Won't Speak It!

God
Delivers Touch for Healing
Whispers Courage For Victory
Anoints Gift For Callling

To Your Spirit!

The ...
Cause for the effects
commands controlling outcomes

When You Seek God's Goodbreak
Christ Sends The ZipFlie Thru Holy Spirit
Pray & Praise Unzips The Touch Downloading The
Goodbreak Commands To Your Spirit

Then You Speak Victory
The_nYou Speak Healing

FORCES OF OPPOSITION
NEED DEACTIVATION
BEFORE ACTIVATION
A MIND HIBERNATION
ON LIVING TRANSLATION
GIVING ANY ADULATION
NOT GOD'S RECOGNITION
IN YOU, GOD'S CREATION

COMPUTE
DELAY

By the word of the Lord were the heavens made; and all the host of them by the breath of his mouth - Psalm 33:6

Be Still, And Know That I am God
BE STILL
PSALM 46:10

SERIOUSLY, COMPUTE DELAY!

Take moaning & bitterness *seriously* not to miss your destiny that has your best overflowing, overfilled abundant life ! God is more interested than yourself in you reaching your divinely touched best, be in agreement! Mistreatment, lost item, damaged stuff, cut traffic, slipped mistake...whatever in the valley, only prepares you, works for you instead of hindering you or working against you! Keep pressing on your built-in pedal: good attitude! All is well! Nothing happens randomly! God may be quiet in the valley, but when you acknowledge He built a good attitude in you and make God the God also in the valley as your promised upcoming mountain top, He will fulfill his promise just like as He said let there be light in Genesis, now be joyful to receive this promise "that your latter days will be better than your former days"!

COMPUTE
OPACITY

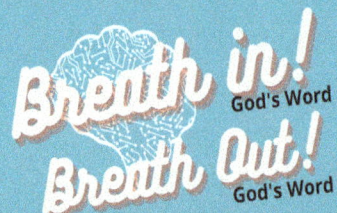

By the word of the Lord were the heavens made; and all the host of them by the breath of his mouth - Psalm 33:6

Be Still, And Know That I am God

BE STILL

PSALM 46:10

COMPUTE

YOUR OPACITY

People can see your guard, adjusted by your faculty of spirit ! And most importantly by whom you guard It, in your spirit's subscription! Your spirit Is your single point of failure or a miracle halo. Our believing in spirit interacts to the geometrical space of our spiritual realm that lets divine sounds whisper (as Moses), walk on water (as Peter), telepathically be met (as Ethiopian eunuch), beyond imparting healing touch & words!

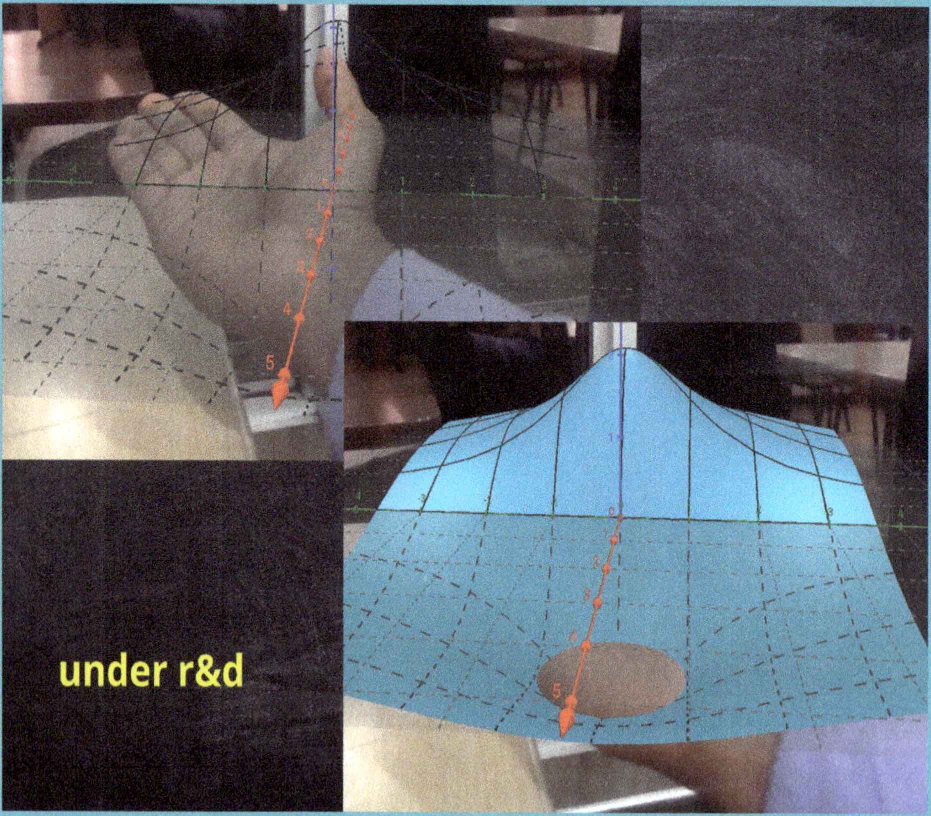

under r&d

COMPUTE
SPRIT THEORY

Breath in! God's Word
Breath Out! God's Word

By the word of the Lord were the heavens made; and all the host of them by the breath of his mouth - Psalm 33:6

Be Still, And Know That I am God
BE STILL
PSALM 46:10

50

under r&d

RESOURCE 3
COMPUTE

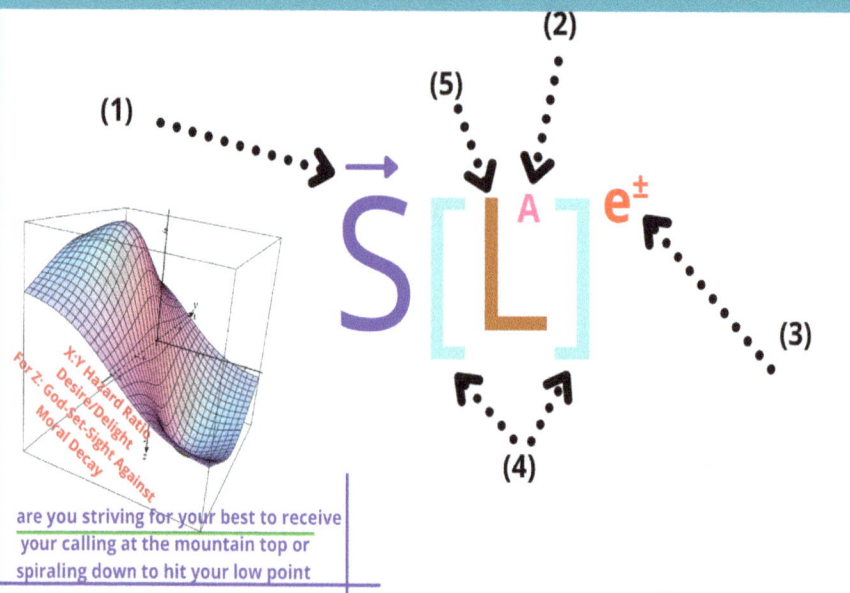

Spirit Theory affirms mankind has a <u>directional attitude</u>(1) with a <u>renewable energy</u>(2) and <u>transferable energy</u>(3) in every <u>moment</u> <u>or state</u>(4) utilized to attain <u>levels</u>(5) of life pursuits.

WORKSHEET 1
INTREGRATE

Breath in! God's Word
Breath Out! God's Word

By the word of the Lord were the heavens made; and all the host of them by the breath of his mouth - Psalm 33:6

Be Still, And Know That I am God
BE STILL
PSALM 46:10

RESOURCE 4
INTEGRATE

Your Word

Call it in, It's A Believing in The Now **Subscription** of Your Sprit in **God's Power**

4 Putting God 1st
4 Healing
4 Surrender
4 Joy

Write /Declare it like this
I call in putting God 1st in all my ways
I call in for healing in my body
I call in for joy ....keep going!

Your Creations

Blend it in, In The Design!

The Product, Design, Result

God's Word

your gift, profession, your seed

WORKSHEET 2
PUBLISH
FOR WHAT YOU BELIEVE
HAPPENING IN LIFE NOW
HAPPENING AT WORK TODAY

Breath in! God's Word
Breath Out! God's Word

By the word of the Lord were the heavens made; and all the host of them by the breath of his mouth - Psalm 33:6

Be Still, And Know That I am God
BE STILL
PSALM 46:10

RESOURCE 5
PUBLISH
Believe Like 2nd Corinthians 1:20 > Yes & Amen

Embed Your Faith, Your Hope, Your Believing, In Your Doings
When You're Hopeful, You'e In A Difficulty

prayer (S-Type)
praise (T-Type)
mercy (B-Type)
inside wire

Believe it like this

Believing for > fruitfullness

It is written.. (Bible Verse) (P-Type Wiring)

e.g It is written, others, like seed sown on good soil, hear the word, accept it, and produce a crop--some thirty, some sixy, some a hundred times of what was sown

Suely, I Believe... (B-Type Wiring)

e.g Suerly I believe, I am a shifter, becuase I have accepted you to be my God pursuing first in all my ways !

Lord, Thank You... (T-Type Wiring)

e.g Lord Thank you I will be fruitful setting a generational blessing at home and at workplace shining bright a light of hope on others with your grace on me

WORKSHEET 3
DAY
BIG PICTURE STRUCTURE

By the word of the Lord were the heavens made; and all the host of them by the breath of his mouth - Psalm 33:6

Be Still, And Know That I am God

BE STILL

PSALM 46:10

DAY

6 Mindful Organizations

#1 MEANING OF YOUR LIFE TO BE PRESENT NOW, TODAY & AFTER { See Page 12, Wire Type: Humility; To Connect & Listen Need }

Goal Setting Prayers

1 - Drift (Struggle State)
e.g. Be humble, people struggle from drift down to praising. Work is here. Ture religion (James 1:27) is here. Giving hands!

2 - Shift (From Vegitate State)
e.g. I declare to surrender, I declare to repent, I declare to dwell on the Word of God in humility, your spirit's subscription

3 - Lift (In Declared State)
e.g. I pray you fight my giants, overcome my opposition, clear my pathway

4 - Touch (Faithful State)
e.g. I will be disciplined to excel at my believing, I will be increasing trust in my planning to be STILL, & Obedient

5 - Honor (Fruitful State)
e.g. I will remember God's goodness, give love, do with kindness, excel for honoring God's gift of life

6 - Praise (Obedient State)
e.g. I will be in good attitude in ups and downs, I will praise, praise, praise consistently in thankful steps & still believings stops

DAY
DO EACH CLUSTER

By the word of the Lord were the heavens made; and all the host of them by the breath of his mouth - Psalm 33:6

Be Still, And Know That I am God

BE STILL

PSALM 46:10

DAY

6 Mindful Organizations
Activities

> #1 MEANING OF YOUR LIFE TO BE PRESENT NOW, TODAY & AFTER { See Page 12, Wire Type: Humility; To Connect & Listen Need }

1 - Drift (Struggle State)
- READ FRUITFUL SPIRIT: GRACE CODE OF CONDUCT
- READ MARK 4:14
- ASK WHO NEEDS HELP IN CARRYING OUT 1-6 DAY ACTIVITIES
- PRAY FOR ONE ANOTHER USING #3

2 - Shift (From Vegitate State)
- Read Bible & Pick A Verse To Recite Through Out The Day
- Read Planner, Listen Message, Listen Worship

3 - Lift (In Declared State)
- Write Out, Your Sustaining Faith (It's Written, I Believe, Thank You) : a) For The Season b) for the day c) for people in drift

4 - Touch (Faithful State)
- Recieve in you Spirit, Your Delivering Faith! Say It / Write out/ Remind At Your In The Moment Need To Oppose Gaslighting Feed (I Declare/I Call In vitality, healing, clarity) ... Ship It, What You Affirmed @ #3!

5 - Honor (Fruitful State)
- Under Your Breath Instant Declaration Now & Then To Recieve To Your Spirit & show love, kindness--excelling for honoring God's gift of life

6 - Praise (Obedient State)
- praise, praise, praise for thankful steps & still believing stops, for what didn't happen, & that happened

DAY
FRUITFUL SPIRIT
GRACE CODE OF CONDUCT

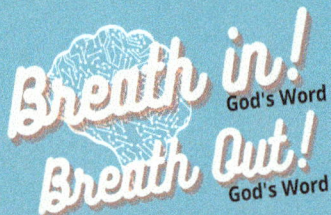

By the word of the Lord were the heavens made; and all the host of them by the breath of his mouth - Psalm 33:6

Be Still, And Know That I am God
BE STILL
PSALM 46:10

FRUITFUL
SPIRIT
GRACE CODE OF CONDUCT

1 - WEEPING
GOD IS A WEEPING GOD (JOHN 11:35) OVER OUR LOSS, OUR DRIFT FROM A GIFT OF LIFE; YOUR LIVING TRANSLATION (A PUTTING GOD FIRST LIFE) HAS A NO JOKE EXPECTATION FROM GOD IN LIFTING ONE ANOTHER (MATTHEW 25:14) LIKE AN EARTHLY FATHER WHO DIES DAILY WEEPING OVER HIS LOST CHILD HOMECOMING (MATTHEW 9 : 35)

2 - STAY
YOU CAN'T AFFORD LOOSING YOUR DELIGHT LANE FOR DESIRE (EPHESIANS 5) FOR THE WORK OF LIFTING ONE ANOTHER IN CHRIST GRACE IS TOO MANY AND WORKERS ARE TOO FEW (MATTHEW 9:36). IN YOUR HOME, YOUR FRIEND, COWORKER, WHO SEEK A DAILY LIFT, YOUR BECOMING

3 - SIGHT
YOUR BODY CAN INSTANTLY TELL WHEN YOU ARE AT DESIRE STRENGTH vs DELIGHT STRENGTH (EPHESIANS 5). YOU CANT SERVE BOTH MASTERS, BUT STRIVE AT DELIGHT. DESIRE GIVES A BURNING DESIRE. DELIGHT GIVES A RIVER OF LIFE SHINING BRIGHT. DESIRE EATS, DELIGHT NOURISHES YOU BODY

4 - WAVELENGTH
YOU CAN'T DO IT LIKE THE OLD DAYS! WHEN YOU ARE DOING YOUR LIVING TRANSLATION — MIND YOURSELF WHO ARE IN THE CONNECTED WAVELENGTH OF DOING : YOUR ADVOCATE, YOUR SPIRIT COUNSEL! LIKE ANGELS WAITING ON GOD, WAIT ON ANGELS. SHOW ANGELS & THIER COUNSEL A RESPECT (JOHN 16:14) !

5 - SACRIFICE
THE MOUNTAIN TOP YOU CAME INTO IS WITH A SACRIFICE FROM MANY MANY MANY GOD'S MERCIFUL SACRIFICE. BE JOYFUL FOR BEING CHOSEN IN YOUR GIFT TO CARRY OUT GOD'S SOVEREIGN PLAN

6 - COHERENCE
ALWAYS CHECK FOR COHERENCE. YOU CAN TELL WHEN YOUR SPIRIT IS WIRED UP TO YOUR MIND FOR DELIGHT GUSHINUP GRACE FROM TREE OF LIFE, AS MUCH AS THE OPPOSITE WAY TO YOUR BURNING DESIRES FLARING UP. DO INSTANT RECITE TO REALIGN, ANYWHERE YOU ARE, THROUGHOUT THE DAY, ASIDE FROM DOING DAY WORKSHEET OF SETTING GOD1ST IN BELIEVING

WORKSHEET 4
NOW
**UNDER YOUR BREATH.
RECITE. RECITE. RECITE**

By the word of the Lord were the heavens made; and all the host of them by the breath of his mouth - Psalm 33:6

Be Still, And Know That I am God
BE STILL
PSALM 46:10

INSTANT
NOW
under your breath declaration

1 - God 1st
I will put God first to be still for obedience

2 - Repent
I repent for my dishonor

3 - Forgive
I will forgive to be forgiven in love

4 - Joyful
I will be at joy with the peace in Jesus Christ (John 16:33)

5 - Surrender
I surrender my will with humility for Holy Spirit

6 - Fruitful
I will be fruitful 100x fold in my ministry to another

7 - Remember
I will remember God's goodness of His mercy for His Glory

WORKSHEET 5
LIFETIME
**BIG PICTURE STRUCTURE
PS: PARENTS DONT MISS FORMATION
EPHESIANS 6:4**

By the word of the Lord were the heavens made; and all the host of them by the breath of his mouth - Psalm 33:6

Be Still, And Know That I am God
BE STILL
PSALM 46:10

LIFETIME
growing in glowing

1 - Formation
Dwelling in the Word of God, Wonderfully to Learn From K-12

2 - Preparation
Cultivate your gift, get mentored

3 - Abundance
Fruitful 100x as in Mark 4;20 lifting, shifting others by your becoming

4 - Remember
Remembering God's goodness for setting generational blessings, setting grace, coaching, mentoring others in their formation, preparation, or abundance

COHERENCE

 God's Word / God's Word

66

By the word of the Lord were the heavens made; and all the host of them by the breath of his mouth - Psalm 33:6

Be Still, And Know That I am God
BE STILL
PSALM 46:10

CONHERENCE
DOUBLE CHECK

INSTANT VIEW
DAY TIME VIEW
LIFETIME VIEW
SHOULD NOT CONFLICT

THERE IS POWER IN MEMORY FOR A COHERENCE WITH YOUR DESTINY! RECITE THE INSTANT VIEW UNDER YOUR BREATH FROM YOUR MEMORY ON THE GO! SEE PAGE 62

1 - Instantly What You Say, The Worm Eye View
2 - Connected In Your Day View,
3 - To Your Liftime Bird Eye View

GIFT
A WELLNESS TO MATURE SPIRITUALLY

By the word of the Lord were the heavens made; and all the host of them by the breath of his mouth - Psalm 33:6

RESOURCE 6

GIFT

FOR WELNESS

Your gift, your talent, your seed is the single most important element stored in you to grow spiritually mature--its the centerpeice for your wellness! It's where the relationship between your spirit and mind is forged! Your gift comes from the command of your spirit to your mind. Now imagine not using your gift, your mind is not at maximum capacity controlling its environment! Your mind is for control, your spirit for command! Between the two, your gift becomes a reason for relationship, with yourself! Your gift, releases the grace found in your spirit, enlargening your joy!

READ ME
BE STILL

By the word of the Lord were the heavens made; and all the host of them by the breath of his mouth - Psalm 33:6

Be Still, And Know That I am God

BE STILL

PSALM 46:10

BE STILL

BE CHILL

1st

The moment you say God 1st, don't be surprised to be led by the Spirit of The Lord, to face up the opposition in the valley (as Jesus Did, Matt 4:1) ! This time not to deal the opposition (as it's already dealt by Jesus), but ONLY to show up & embarrass the opposition at this check point & to collect you visa stamp which has your name on it for the crown of life (James 1:12)! All the spamming is to discourage your steps of faith!

FAQ

By the word of the Lord were the heavens made; and all the host of them by the breath of his mouth - Psalm 33:6

Be Still, And Know That I am God
BE STILL
PSALM 46:10

FAQ

1. Getting Bitter Randomly? Refocus on your shift dwelling on God's Word, or listen to worship songs (breath in fix)! Refocus on your lift, through your gift if you are in a position of pursuing it in the season of life you are in, refocus on the next best that is in God's hand, which has yours & your family promotion!

2. Loosing Peace to Circumstances? Without testing God, without necessarily fighting it but connecting to it being still, and praying that God is behind every circumstance and letting God fight your battels, while you keep focusing on manifesting good attitude with out denying the fact

CLOSING

By the word of the Lord were the heavens made; and all the host of them by the breath of his mouth - Psalm 33:6

Be Still, And Know That I am God
BE STILL
PSALM 46:10

I AM
EXODUS 3:14

Rock Bottom

The moment you hit a rock bottom AND you start praying, God becomes the rock you stand on, turning it around like a Holy Ground to meet you as he did with Moses: speaking, answering to your prayers right there & then!

By the word of the Lord were the heavens made; and all the host of them by the breath of his mouth - Psalm 33:6

Be Still, And Know That I am God
BE STILL
PSALM 46:10

75

WEEPING
JOHN 11:35

God is willing to sow in & show out – out of season blessings! Weeping! To the point of giving up His generations stored in showtime miracle : resurrection! Resurrting Lazarus of Bethany! The first in line miracle set for the Son of God! But God! But God who was dear & near, not to Lazarus of Bethany, but to his Sister Mary, did it to a best friend of friend ! That far! But God!

Why Be Still With God's Word

Imagine doing a shopping without a cart, and you carry the watermelon 🍉, the freezing ice cream! That's what's like to living your day without carrying the Word of God! Heavy to get ya weary with all emotional ups and downs you go through! Giving or receiving a smile consistently in the middle of this shopping with all the warm, cold and heavy lifting without a cart isn't possible! So is not without the Word! A cart lets you push through not carry the burden, so does the Word! Try it!

www.ingramcontent.com/pod-product-compliance
Lightning Source LLC
Chambersburg PA
CBHW062042290426
44109CB00026B/2709